I0128908

LONG AND LOW

"So much of the material in this book is completely enthralling. I need to have it on my shelf so I can read, then re-read it."

Anne Wilson, author of *Unity between Horse and Rider* and classical riding teacher

This short book examines the history, origins, and unintended consequences of the commonly used and widely accepted practice of "long and low" riding and training in dressage. International rider, trainer, writer, and teacher Paul Belasik explains the little-known origins of the mysterious practice that has evolved into a major modern-day controversy.

Long and low riding has negative effects on the quality of dressage at the highest levels of the sport and widespread confusion for riders of all levels. This important book seeks to educate riders on why it is taught, where it came from, and how the unintended consequences have played out for riders all over the world. It is essential reading for all dressage riders, and any event riders and jumpers wishing to better understand the origins of long and low.

LONG AND LOW

A Revolution in Modern Dressage

Paul Belasik

CRC Press
Taylor & Francis Group
Boca Raton London New York

CRC Press is an imprint of the
Taylor & Francis Group, an **informa** business

Designed Cover Image: Images by Diana Mercean

First edition published 2026
by CRC Press
2385 NW Executive Center Drive, Suite 320, Boca Raton FL 33431

and by CRC Press
4 Park Square, Milton Park, Abingdon, Oxon, OX14 4RN

CRC Press is an imprint of Taylor & Francis Group, LLC

© 2026 Paul Belasik

Reasonable efforts have been made to publish reliable data and
information, but the author and publisher cannot assume responsibility
for the validity of all materials or the consequences of their use. The
authors and publishers have attempted to trace the copyright holders
of all material reproduced in this publication and apologize to copyright
holders if permission to publish in this form has not been obtained. If
any copyright material has not been acknowledged please write and let
us know so we may rectify in any future reprint.

Except as permitted under U.S. Copyright Law, no part of this book
may be reprinted, reproduced, transmitted, or utilized in any form by
any electronic, mechanical, or other means, now known or hereafter
invented, including photocopying, microfilming, and recording,
or in any information storage or retrieval system, without written
permission from the publishers.

For permission to photocopy or use material electronically from this
work, access www.copyright.com or contact the Copyright Clearance
Center, Inc. (CCC), 222 Rosewood Drive, Danvers, MA 01923,
978-750-8400. For works that are not available on CCC please
contact mpkbookspermissions@tandf.co.uk

Trademark notice: Product or corporate names may be trademarks
or registered trademarks and are used only for identification and
explanation without intent to infringe.

ISBN: 978-1-041-04260-0 (hbk)
ISBN: 978-1-041-04258-7 (pbk)
ISBN: 978-1-003-62757-9 (ebk)

DOI: 10.1201/9781003627579

Typeset in Joanna
by Apex CoVantage, LLC

CONTENTS

ACKNOWLEDGMENTS

I would like to thank Anne Wilson for initially suggesting Taylor and Francis as a possible publisher for this most recent book of mine. I am also grateful to Alice Oven at Taylor and Francis for her encouragement and support in getting the new book out, as well as the editorial and production teams who went over the book so carefully. This book would not be possible without the copious editing and manuscript preparations of my wife Rose Caslar Belasik. I would like to give a special thanks to Diana Mercean for her excellent illustrations. I would also like to thank Dr. Hilary Clayton for allowing me to use illustrations from her book, *The Dynamic Horse*.

ABOUT THE AUTHOR

Paul Belasik is a highly respected international rider, trainer, writer, and teacher. An avowed proponent of classical equestrian ideals, he has published eight books on dressage and riding, including the perennial favorite *Riding towards the Light*, and most recently, *Dressage for No Country*.

A graduate of Cornell University, Belasik has ridden and trained at every level in dressage, from young horses to beyond Grand Prix into airs above the ground. He has successfully competed at the highest levels in competition. He also has extensive experience in training and competing through the upper levels of eventing, which encompassed the early part of his career before he turned solely to his first and true love of classical dressage. Belasik has sought wisdom from great riding masters such as Dr. HLM van Schaik and Nuno Oliveira, and his wide-ranging studies include the concepts of Zen Buddhism and martial arts. Paul has also been a lifelong breeder of horses, first with Trakehners and then Andalusians. Paul's Andalusian stallion Excelso successfully competed in international Grand Prix competition and sired a number of upper-level horses, including offspring which have competed through Grand Prix.

Paul Belasik gives clinics, lectures, and demonstrations internationally. He trains a wide cross section of clients at his Pennsylvania Riding Academy at Lost Hollow Farm, where his short courses concentrating on the rider's position have brought him acclaim from students who have come to the farm from all over the world.

Belasik has helped all levels of riders from around the world for over 40 years. They are a diverse group: members of various national and international equestrian teams, a North American Endurance Champion, Handicapped champions, and riders of various levels with no interest in competition. His training methods focus not only on the practical, physical point of view, but also keep a keen eye toward the artistic, scientific, and philosophical components of horsemanship. In 2002, Belasik produced a major work, *Dressage for the 21st Century*. This book covers all aspects of riding and training and marries a genuine adherence to classical principles with modern thinking.

His website is https://paulbelasik.com/index.php/paul-belasik/

INTRODUCTION

Early 1800s

I have in front of me a large book: *The Horse: 30,000 Years of the Horse in Art*, by Tamsin Pickerel. In it there are copies of primitive stone paintings, there are images of sculptures, tapestries, paintings, and portraits. In all the depictions of a human being riding a horse throughout the history of humans on horseback, there is not one where the rider is deliberately riding the horse with its head down. It had certainly never been done in the history of

DOI: 10.1201/9781003627579-1

dressage. Starting with the flourishing of the Renaissance some 500 years ago, a consistent lineage of training originated in the schools of Grisone, Pignatelli, and Fiaschi in northern Italy. Noblemen from all over Europe came to these schools to study and returned to their native countries: Vargas to Spain, Pluvinel to France, Lohneyssen to Germany, Antoine de Pluvinel to England. They taught what they learned to the next generation of William Cavendish in England, Guérinière in France, and so on. They practiced an art form, an equestrian dance. They were obsessed with the airs above the ground. The Duke of Newcastle stated in the 1600s that the whole object of the school was to get the horse upon the haunches. They became masters of collection. In Hans Handler's book on the Spanish Riding School, which chronicles the training of the SRS for 400 years, he includes an elegantly simple page of line drawings of horses showing the evolution of the dressage horses training. It begins with the horse in a natural state, with the balance tipped to the front and goes on to show, as the training progresses, the balance being moved farther back over the powerful hindquarters. One can see the progression lightening the forehand and strengthening the rear until a rider and horse could rear or jump off the hind legs in perfect control, mastering collection.

In the 1800s, a brilliant rider named Gustav Steinbrecht trained horses in Germany. In his famous book *The Gymnasium of the Horse*, he wrote:

> I have had horses whose front legs were completely worn out, but whom nature had endowed

with strong hindquarters and I have given them such complete freedom in the shoulder and such reliable gaits by bending their haunches that they could compete with the best and most valuable of their species. Under former riders their strong hindquarters had merely helped push all the weight toward the forehand with great force, with such overloading ruining the forelegs in a short time. By taking the load from the forehand, these horses gradually regained their natural elasticity and agility. The higher elevation of the neck also gave them increased action from the shoulders.

This was the kind of thinking that evolved for some 500 years, when the modern art of dressage began at the same time as human ballet. It was an art form and it was based on the biomechanics of being able to change the horse's natural balance on the forehand toward the powerful hindquarters which could be accessed to perform dance-like movements. Then in the 1800s, around the same time Steinbrecht was writing, a system of riding and training horses that had never been done before began to appear. Its ramifications are still having an effect on riders and horses in the present. It was the beginning of riding the horse with its head and neck down. It was called "long and low."

1

THE BIRTH OF LONG AND LOW

GERMANY 1832–1852

Louis Seeger was very worried. Seeger was trained under Maxmillian von Weyrother at the Spanish Riding School; he was steeped in the principles of Guérinière. He was now a very influential trainer in Germany. Seeger was an avowed classicist.

Seeger was concerned about Germany's dressage riding falling away from the classical principles. "Our army's riding system which was introduced after the Peace Treaty of 1815 already had many similarities to [Francois] Baucher's methodology" (Seeger 83). Perhaps Francois Baucher's

DOI: 10.1201/9781003627579-2

most significant contribution in this foment of the times was his disdain for classical collection. He claimed that he alone understood the dynamics of dressage, and the lynchpin of his theory was that the horse needed to balance over four legs. It should move more like a spider.

In his time, Baucher was a darling of the glitterati. His circus shows were very popular, he was a star showman. Seeger had not only seen his performances, but he also read his book and he had at Baucher's invitation a chance to actually ride his horses and witness the training, including severe manipulations of the jaw and neck. Seeger was aghast. Baucher's words and writing on lightness did not in any way match his riding. Seeger felt he had to do something. He wrote his own book, a scathing rebuttal of Baucher's technique: *A Serious Word with Germany's Riders*.

Seeger described Baucher's use of the bit:

> This bridle works the same for Baucher, even on the greenest horses. To encourage the yielding of the jaw at the beginning of the horse's training, he pulls mainly on alternating reins of the Weymouth (for the purpose of the schooling, more than on the snaffle reins). He uses the shanks to flex the jaw and to force the head and neck from side to side and down.
>
> (12)

Seeger wrote emphatically in his conclusion:

> There is only one way to get out of the labyrinth the art of riding has gotten into. This is to first ride the

> horse in the natural position in order to develop
> the impulsion. Next, one should raise the head and
> neck to develop shoulder freedom and the bend-
> ing in the stifles. From this position, one should
> perfect the bending of the haunches so that the
> movement of the horse comes closer to the school
> movement. This will make the horse solid and agile
> at the same time.
>
> (84)

If anyone heard Seeger, it was Gustav Steinbrecht. Steinbrecht was a student of Seeger. Steinbrecht went on to write *The Gymnasium of the Horse*, which would later be referred to as the German dressage bible. Steinbrecht was a brilliant horseman. Steinbrecht echoes Seeger's admiration for the classicists. He eloquently explained the advantages of that dressage training in practical, clear language that stands up today. However, there is an old saying that a prophet is the man in the next county. What was going to stamp the modern dressage in Germany?

At the time of writing his book *Classical Dressage Training in Practice According to the H. Dv. 12*, Gert Schwabl von Gordon was by his own description the last living representative of the Cavalry School of Hanover, Germany. He felt obliged, he said, "to elucidate the principles of the classical art of riding as they were understood and cultivated at the world renowned school" (9). However, as we'll see, they had redefined "classical." Schwabl von Gordon presents us with a unique eye-witness account of the birthplace of long and low. He goes on to explain how the north star of the training at the Calvary School of Hanover was the H. Dv. 12.

In the preface to the English edition of H. Dv. 12, publisher Richard Williams writes, "H. Dv. 12 has its roots in
the regulations dating back from 1882. These regulations
summarized the knowledge gathered together in cavalry
training since the 18th century" (ix). The H. Dv. 12 was
the German Cavalry Manual on Training the Horse and
Rider: Army Regulations 12. These regulations are brief
in their guidelines. There is not a lot of information or
explanation of long and low or how to achieve it, and
what is there can appear to be contradictory. In the manual under the section *Part C. Schooling of Horses, XI. Dressage
Horses during the First and Second Year*, it goes on to talk about
looseness (losgelassenheit). It says:

> The horse must learn to regain the posture that it
> has found without rider and move just as uncon-
> strained—with a long neck and a low nose—under
> the weight of the rider. If it is able to maintain this
> unconstrained movement, it is "loose."
>
> (98)

Finally, it is stated in bold letters, **"Looseness [losgelas-
senheit] of the horse is the basic precondition for the
success of the entire dressage"** (98). It is interesting how
nonchalantly this revolutionary idea, that a horse's natural
posture in movement "with a long neck and a low nose,"
is laid out for all of the riders of the German cavalry. It is
an order issued from regulations to be obeyed, that this
posture is a precondition for the entire success of dressage.

It was clear that among a worldwide community of
classical dressage trainers, this system of long and low
was an interpretation and a product of a very specific

subset of instructor/trainers, mainly from the German military and the Cavalry School of Hanover in particular. In his book, *The Art of Dressage*, Alois Podhajsky quotes from the book, *The Noble Horse*, which is by German rider Baron G. Biel, published in Dresden in 1830:

> The author begins a chapter with the statement that a sad state of affairs prevailed in most German Riding Schools, that the horses were overbent, their heads pulled down in an unnatural manner, and that one sought in vain for an example of the true art of riding. He [Biel] continues literally: "I came to Vienna and watched the riders of the Spanish Riding School, I felt that riding was practiced on different principles than in those riding academies I had visited previously."
>
> (Podhajsky 20)

Long and low was not practiced at the Spanish Riding School in Austria, nor academies in France, Spain, Portugal, or England.

What did this long and low look like and what was it supposed to do? According to Schwabl von Gordon:

> Here is a summary of the most essential features of riding long and low. Two equally important criteria should be emphasized here: stretching the horse's neck and dropping it down while yielding the poll with the nose of the horse on the vertical. The horse shouldn't loll around on the bit—ie, the rider shouldn't allow the horse to have a fifth leg. Instead, there should be a light but constant

> contact between the rider's hands and the horse's mouth. To designate the point of the shoulder as the fixed point of reference for lowering the neck is erroneous because only by really stretching downward as though the horse is tracking something or looking for truffles, can the rider ensure that the horse, its back and the broad muscle (latissimus dorsi) is stretching as well.
>
> (17)

He goes on to say the engagement of the hindquarters is not the objective, only suppleness. This suppleness will somehow "create the conditions necessary for engagement of the hindquarters in the working frame" (17). There is no explanation for how this is supposed to evolve into collection.

However, to be fair, there is cursory acknowledgment that this all be done with a light rein. There is also instruction that the horse should "immediately comply with the request: 'From the outside regulating rein, have the horse stretch into the long and low position!'" (Schwabl von Gordon 18). It does not take much imagination to see how the scene Baron Biel describes of riders in the German schools riding with horses "overbent, their heads pulled down in an unnatural manner," could have developed with the chief instructors teaching this system of immediate compliance of the horse to lower its head (Podhajsky, 20).

In its simplest description, in this system the horse is trained to immediately drop its head and neck to a position much lower than the shoulder, "searching for truffles," when the rider uses the reins, particularly the outside rein, amounting to redistributing more weight onto the front legs and shoulders. The automatic disengagement of the hindquarters is never addressed. The implication is that the value of the stretching of the back is so important that it supersedes all the unintended consequences. We see here the seeds of a tension that will appear and reappear, between, on one hand, an idea of stretching of the horse and, on the other hand, an understated, almost unwritten insistence on immediate submission.

There are notations in Schwabl von Gordon's book that this new long and low system of the Hanover School was supported by Gustav Steinbrecht and Waldemar Seunig, two very important equestrian authors. I have already quoted Steinbrecht in the introduction regarding the importance he applies to lowering the haunches, that is, redistributing the weight of the horse to the rear. In his masterpiece of writing, *The Gymnasium of the Horse*, Steinbrecht says:

> The old masters knew how to thoroughly work the haunches and have become a lasting example for years to come. Their extraordinary accomplishment which at present are considered to be almost miracles, were a result of this work. . . . *The border-line between military riding and upper level dressage if such a line is drawn is defined in that military riding has its goal to put the horse in balance, that is to put equal weight on the forehand and the hindquarters, while in upper level dressage the movements become*

more perfect the more the center of gravity of the weight can be moved backwards without interfering with the freedom of the hindquarters.

(Emphasis mine, 125)

To me, the further genius of Steinbrecht can be seen in the statement:

There are many riders who are able to speak about bending the haunches and they also believe that they are working the haunches of their horses without ever having felt correctly bent haunches, yet there are enough of them who think the hocks are the haunches, and they have hardly any idea that there are even stronger and more important joints above the hocks.

(124)

It seems impossible that the proponents of the School of Hanover's system, who claimed that long and low was the basic precondition for success of the entire dressage, would call on Gustav Steinbrecht for corroboration when he so clearly expounds on the opposite strategy of dressage training.

As far as corroboration from Waldemar Seunig, in his encyclopedic masterpiece "Horsemanship," there is an index and glossary that goes on for some 28 pages. It covers everything in expansive detail. In the whole work, there is almost nothing about looseness, losgelassenheit, except in its use to stretch at rest. It is certainly never mentioned as a cornerstone of dressage training.

If long and low was so clearly at odds with the classical theory and system, where did this revolutionary new dressage training get its power from?

Long and low's appearance in the late 1800s coincides with the dissolution of the world's cavalries and the birth of competitive dressage. With the advances in mechanized warfare, the cavalry became obsolete. Facing the threat of the loss of their jobs, many of the military horsemen embraced an expansion of riding to include the public. Forward-thinking horsemen like the controversial German Gustav Rau championed the repurposing of cavalry and farm horses to breed modern sport horses. (In a relatively short time, two world wars would decimate Germany. They had little to rejoice in besides their horses.) Rau's greatest skills were probably organizational. Said to be dictatorial, he advanced the idea of German Riding Clubs to interest all German people in the sport of riding and to promote the sport of dressage.

In 1900, the first Olympic Equestrian events were held in Paris. The FEI was formed in 1912. In 1913, Rau was appointed to General Secretary of the German Olympic Committee for Equestrian Sports (DOKR, just before WWII.) In 1938, Rau became the Chief Equerry of Germany and Master of the Horse. Rau's nationalism and racism cannot be swept under a carpet. He was instrumental in trying to install Hanoverians to replace the beloved Lipizzaners at the Spanish Riding School when Germany annexed Austria. Rau ruled German Equestrian Federation politics for 50 years. Later in his life, Rau received an Honorary Doctorate from the University of Bonn. Rau was not an intellectual, but a self-educated man with many untested theories. He was never a dressage rider. When he did talk about riding theory, the language was more filled with ideas of suppleness, not collection. His words echoed the Cavalry at Hanover School's language about looseness. Rau saw the value in terms of publicity from seeing German horses and German riders on top of winner's boxes in the new competitions. He helped create a nursery for sport horses and riders. After some stinging losses for German riders in 1928, he founded a special dressage stable at the Cavalry School of Hanover.

Rau installed Otto Lörke to run the program. Lörke fitted the mold. He learned the German military system when he was serving his time with the Ulan Regiment of the Royal Prussian Guards. Lörke was by all accounts not an elegant rider, but would prove to be good in the new sport of dressage. It was at Hanover that he began to develop a machine producing competition horses and riders that could win.

In his book *My Dancing White Horses*, Alois Podhajsky recounted Lörke offering him a German horse to compete:

> I recall the famous German riding teacher Otto Lörke greeting me at a show in Berlin, immediately after the Anschluss [Germany's annexation of Austria in 1939], with the words that he would have a horse for me in the 1940 Olympic Games that was already good enough to win the top dressage prizes. He simply could not understand my refusal of this offer and my answer that I had never entered anything in competitions except my own work. He, with his typically German mentality,

could not follow my train of thought, and walked
away shaking his head, probably considering my
attitude excessively arrogant.

(57)

Lörke worked with Colonel Baron Holding Berstett who
later became the president of the FEI. In those early days,
the sport was tightly knit. Berstett as president would
have ultimate authority over all the judges of the FEI. This
network would become very important.

It became clear that the marketing experts of the
burgeoning dressage industry were aware of the optics
of German officers riding German horses in a sport of
precision riding. It wasn't long before Locke presented a
team of young women riders, two trained by him and the
third by his student Willie Schultheis. In 1956, Liselott
Lisenhoff, Anneliese Kuppers, and Hannelore Wiegand
brought home a silver medal from the Stockholm
Olympics. The warmblood was now a horse for everyone,
every demographic. With the German horse however,
came an official owner's manual. It was not the manual
of Steinbrecht with his emphasis on the classical elements
of dressage, it was a continuation of the German military
style. Even at the highest levels, the emphasis seemed to
be on the front ends of the horses. Many photographs
and even Schultheis's instructional videos demonstrate
horses disconnected, exaggerated in the front end and
with difficulty engaging the hind end. The rider was not
balanced over the leg, but was heavy-seated.

All these new attitudes concerning the promotion
of dressage as a sport were dominated by military or

ex-military men. Civilians were not even allowed to compete until 1953. The military men who dominated the control of these equestrian events did not foresee any problem in turning the art of dressage into a sport. Unlike showjumping and three-day eventing (which interestingly was simply called Military in the beginning), where it was much easier to calibrate a winner, the subjective element of dressage as in all arts would prove almost impossible to control. The German military equestrians were at the forefront of this new sport. Their philosophy and training system came with them.

> These principles of the classic art of riding were put into practice at the Cavalry School of Hanover and were so successful that Germany's riders from the Calvary School won all six of the Equestrian gold medals in the 1936 Olympic Games in Berlin in individual as well as team competitions.
>
> (Schwabl von Gordon, 12)

In reality, it wasn't the principles that were entirely responsible for the results. Schwabl and the German military had one narrative, but there was another from Podhajsky who was a medalist at these very same Olympics. In *My Dancing White Horses*, he writes:

> [A]fter a long wait, the result of the main dressage event was announced: "Kronos (Germany) first, 15 points; Absinth (Germany), second, 18; Nero (Austria), third, 19; Theresa (Sweden), fourth, 26," and so on. I also noticed all round me too a great deal of head shaking over this result which was

discussed everywhere, including the newspapers. The German judge was responsible for pushing me back into third place, because he not only put his own three countrymen first, second, and third, but placed me only seventh, unlike his four fellow judges, who had all put me somewhere in the first four. In a long article on the decision, the German paper *St. Georg* reckoned that I was entitled to third place even without the Austrian judge, who had placed me first. Talking of this disappointment, it is not without interest that many years later, in 1943, the influential secretary of the German judge announced openly during the lunch at the Jockey Club in Vienna that he had deliberately marked me down in the 1936 Olympics, being determined that a German rider should win. This meant marking up his countrymen and downgrading the dangerous favorites, and his tactics cost me at least the Silver medal, if not the Gold in Berlin.

(42)

From the beginnings, the sport was consumed with politics and intrigue. The German horse industry continued to expand and what helped was a corporate culture. Fritz Schilke, the father of the modern Trakhener, was famous for his advice that "no breed ever died from lack of type, breeds die from lack of marketability." These warmblood horses were being imported all over the world with a multipronged marketing, breeding, and showing strategy. The German philosophy of training these horses to win came right with the horses. It was not the training system of Steinbrecht, it was the training system of the Cavalry School of Hanover. The continuation would also have to involve a complicit cooperation from the judges in this new sport.

The scoring in dressage was so plagued with biases and controversies that in 1952, the International Olympic Committee had decided to ban to dressage from the Olympics. It was only allowed to remain after promises of serious corrections to the system. However, as industry and sport grew, the incentives for biases only became stronger and more valuable. Judges would have to be trained to accept the superiority of the military system and breeders would have to train judges why their horses should win. What could be better than if many of the judges were already breeders?[1] The success of this marketing strategy was proven when much later a scientific study proved the distinct bias for warmblood horses in dressage competitions and a virtual monopoly in showing in the sport of dressage.

Long and low did not get its power from being part of some groundbreaking advance in the 500-year-old history of classical dressage. It got its power from being inextricably linked to the winning successes in the new sport of dressage. Long and low's associated grip on competitive dressage remained to the extent that as of today, it is a required movement in elementary dressage tests in the United States, "the stretching circle," where the horse must immediately comply and lower its head.

Note

1 In the United States in the last 20 years three out of four top dressage judges were also warmblood horse breeders.

2

THE BONES OF THE REVOLUTION

If you are going to ask a horse to carry the weight of a rider and saddle which can amount to an increase of as much as 20% of the horse's weight, it seems only logical and ethical that you prepare the horse to carry the additional weight that lands right in the middle of the suspension bridge of the horse's back. A feral horse or an untrained horse will carry 58% of their weight on the front legs. If the horse standing still raises its neck, even without moving the feet, it will adjust the center of mass and relieve the weight on the forehand by several percent (Figure 2.1).

DOI: 10.1201/9781003627579-3

Figure 2.1 The effect of head and neck position on the location of the horse's center of mass shown by the gray circle for the neutral position. Stretching the head and neck forward moves the center of mass forward (black asterisk on left). Retraction and elevation of the head and neck moves the center of mass backward and upward (black asterisk on right). Used with permission from Dr. Hilary Clayton: "The Dynamic Horse" (134).

If the horse lowers its neck, conversely it will bring the center of mass forward and will increase the weight on the already overburdened forehand by several percent (Figure 2.2).

At first glance, it might seem this positioning of the neck in a high, strong, reaching arch with the poll at the highest point and the face near vertical has a relatively small effect of a few percentage points on changing the longitudinal balance of the horse or affecting its back to help carry the additional weight of the rider. However, the positioning of the neck is critically important, because the neck is the gatekeeper to connecting the topline of the horse. If the topline can be made to be one strongly connected piece, and if the rider can get the horse to start bringing the great muscles of the hind end more under the mass, the rider has the possibility of lightening the forehand, not by a few percentage points

Figure 2.2 The traditional method for starting the dressage horse is by lunging in side reins, without the weight of the rider (Johann Elias Ridinger, 1698–1767).

but by an infinite range all the way up to 100% lightening of the forehand, as in the levade.

The traditional method for starting the dressage horse is by lunging in side reins, without the weight of the rider. The side reins help position the neck either up or down, with more or less bend right or left, while the horse learns to engage the hind end in a series of transitions and changes of bend, all without using the neck as an unridden horse would. Some horses will come to the training with a wonderful sense of balance and will be conformed to use the hind end almost naturally, but for many horses, like people lifting things the wrong way, they will have to be helped. (In a sense, the natural movement of the horse becomes a moot point when you add the unnatural weight of the rider. There must be some ethical preparation.)

The principal aim is to strengthen the neck and develop its connection with the entire topline. By moving the side reins up or down or adjusting the left or right bend, the lunging can help correct abnormalities in conformation to facilitate better balance. Suppose you want to take a photograph of your horse: you whistle, maybe shake a can of grain, throw your hat up in the air. We are looking for that moment when the horse's neck is up in a beautiful crest, ears are pricked, the form is bristling with alert attention. It is a symbol of beauty and power. This is what we look for and try to train for when the horse is on the bit; the poll is at the highest point, the face of the horse near vertical. We are rightly drawn to the beautifully ascending, powerful neck.

One of the main muscles that is principally responsible for the great arch is the splenius. This muscle, of course, does not act alone. It has a very important job of connecting almost the whole topline of the horse. It starts at the shoulders and then connects to the great nuchal ligament system. The nuchal ligament system is a powerful cable that runs the length of the horse's topline, and can adjust tension in it. It has a relationship with the neck vertebrae, and also connects with the large muscles on the underside of the neck which projects the front leg forward in a high lift of passage or maximum reach of an extended trot. It also ties in with the long back muscles. It connects the whole topline and when it braces, it could hollow the topline if it were not for the abdominal muscles which act like the string of a bow and arrow. When we tighten the string, we add more tension to the arching bow. The roundness of the classically trained horse comes from the muscular juxtaposition. The muscles of the topline don't pull against the spinal processes but they pull against the abdominals. This was the old concept of the ring of muscles. When we lunge in this connected, uphill form, we build the strength of the back, the neck, and the whole topline to prepare it to carry weight.

The young horse, of course, does not have the strength to sustain collection, but in a sense we are programming the muscle memory to be in the position to safely carry the weight of the rider and to eventually collect. It becomes a systematic, logical progression. We may start with the side reins long, and we gradually adjust them to bring the poll up or down depending on the horses conformation. And, we can adjust the bend left or right to work on the horse's normal asymmetries, that is, handedness. The side reins connected to the bit shape the neck gradually like braces on a child's teeth. The horse learns quickly to relieve the pressure of the bit by flexing its neck. The misunderstanding is the side reins are using force and make the horse lean on them. Horses are very smart. Especially since there is no rider on top, they can have immediate feedback on the self-created pressure. Suppose you have finished riding, you're in a room standing near a chair. You are trying to change out of your breeches. You balance on one leg, but the pants are tight and you struggle to get the other leg out. You begin to wobble so that you get close to the chair and just touch your hip against it. You barely have any pressure on it at all but the propriocentric system in your body uses the touch for a reference point and you easily get the leg out with perfect balance. You are not using the chair to hold on, you barely touched it with your hip or maybe your elbow. The important thing is that you didn't restore your balance by leaning on the chair with all your strength. This is the secret of how properly used side reins work in blocking the neck from leaning down. The horse develops its back and hind end to balance. The genius of this classical system is that it shapes the neck gradually until the horse gets strong enough to balance and rebalance by merely touching the bit. The bit is a propriocentric reference point for the horse's whole field of longitudinal balance. The strength comes from a coordinated effort of the whole body, not from using the contact as a crutch.

Figure 2.3 If you lunge with the neck down, the strengthening of the balancing muscles, tendons, ligaments, and powerful hind-quarters does not occur. The bow of the bow and arrow gets its support more from the stretched nuchal ligament, and you add even more weight to the already overburdened front legs.

This only works because in proper lunging, we develop the balancing muscles, tendons, ligaments, and powerful hindquarters. Now when we do add the weight of the rider, the horse had been prepared. On the other hand, if you lunge with the neck down, none of this strengthening occurs. The bow of the bow and arrow gets its support more from the stretched nuchal ligament, and you add even more weight to the already overburdened front legs (Figure 2.3).

The position of the neck unfortunately has such great influence because it can be the "spoiler" to this systematically trained weight shift. If the neck is allowed or trained to be loose, long, and low, it will always be a potential escape for collection, not giving the hind end and back anything to connect with, and adding more weight to lift. These horses cannot develop enough core strength for leverage. A pry bar made of rubber is useless.

Imagine you are in the gym and are going to lift a barbell. You step up to the weight, flex your knees, and straighten your back, but the trainer rolls the barbell away from you by a foot or so. If you try to lift without fixing your stance, what do you think is going to happen to your lower back? Unlike the feral horse or an unridden horse, the ridden horse must be trained not to use its neck for leverage. The ridden horse must stabilize the neck, taking it out of the balancing equation, and instead use the muscles of the core, back, and hind end to come under the weight the same way the human needs to be trained to block the instinct of lifting when stretched out, but instead firm and straighten the back, bend the knees, and come under the weight. An interesting note can be made here: one of the worst human jobs for back problems and pain is that of a chambermaid, a person who makes beds all day long, often carrying nothing much heavier than a sheet or blanket, but constantly locking the knees and leaning over, stretching while lifting, day after day. Even without much weight, this posture is harmful (Figure 2.4).

The two opposing ideas about how to train the necks of dressage horses would be a source of conflict for a hundred years. The classicist's answer was to stabilize the neck, to make a strong connection of the topline, and train the back end of the horse to be strong, flexible. This position could sustainably carry the weight of the rider and eventually master the balance of collection.

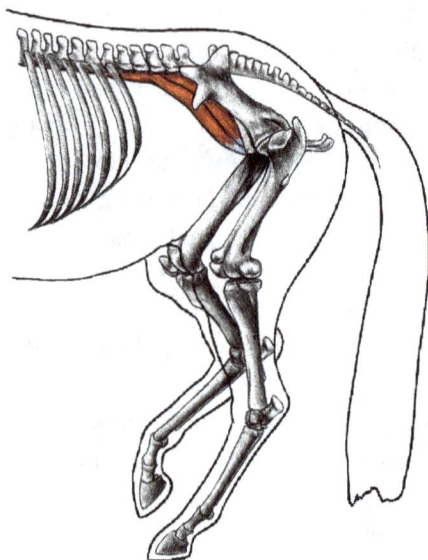

Figure 2.4 Out of sight, deep beneath the skin lie the psoas/ iliopsoas group of muscles. They are responsible for bringing the hind end down and under the body. They are some of the most important muscles in all riding, but especially in dressage and collection.

In the long and low system, the object was to completely subjugate the neck, flex it, bend it, make it completely submissive to the pressure and control of the bit. This submission would theoretically improve obedience and this obedience would improve performance.

The First Fulcrum

Out of sight, deep beneath the skin lie the psoas/iliopsoas group of muscles. They are responsible for bringing the hind end down and under the body. They are some of the most important muscles in all riding, but especially in dressage and for collection (Figure 2.5).

If the hips and hind legs of the horse were not brought under the mass to help carry, the horse could not be made strong enough to safely sustain upper-level movements with a rider or have any chance of the acrobatic

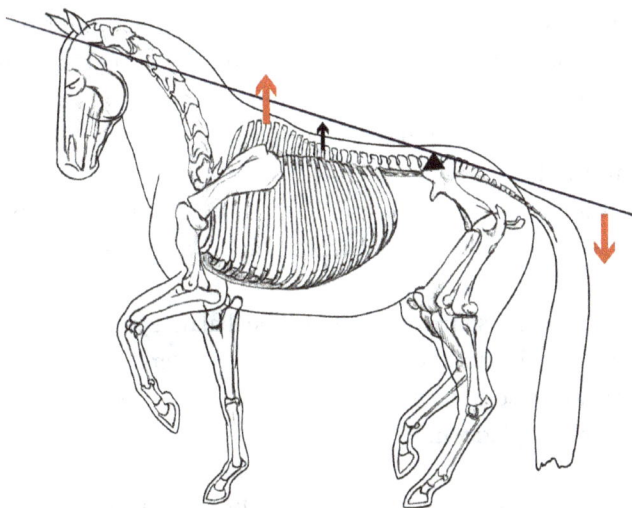

Figure 2.5 In this illustration, the fulcrum is the lumbosacral joint. Training the horse to use this fulcrum ultimately can make the hind end strong enough to control the center of balance so completely that the front end can be entirely lifted off the ground as in levade or airs above the ground. This fulcrum at the lumbosacral area is the fulcrum that classical dressage has always been most concerned with. It is not the fulcrum the German Military riding establishment was most concerned with.

movements of high-school dressage, much less the simplest collection.

In Chapter 1, we saw that Gustav Steinbrecht pointed out a clear difference between horses that just squat down in the hind end and horses that come under and sit in the classical sense. The difference between a horse squatting down in the hind end versus coming under and sitting is that the former may be lowering the haunches, flexing the hocks and stifles, but the hips are out of position. The croup is flat and the hind legs are camped out behind an imaginary vertical line from the point of the buttocks to the ground, the topline is not round. In the latter horse, the haunches are lowered and the hind joints are flexed but the hips have tipped and come under, the hind legs almost seem like they search for more purchase on the footing, as they prepare to carry more weight. An example can be seen in a correct piaffe: the hind legs might widen their stance slightly as the balance shifts more dramatically to the rear. The hindquarters have been developed so strongly that this coming under gesture can serve as a platform to relieve the forehand of literally all its weight as in a levade or a courbette.

This is the same way a weight lifter will widen and settle their stance as close to the center of mass of the weight they are going to lift as possible, in order to use their legs instead of dangerously straining their lower backs. When we see horses during collection with narrow stances in the hind legs or haunches that sway, we know they are not carrying the weight behind and most times—in very complex ways—are propping themselves up with the front legs or creating ground reaction forces that bounces the forehand up. This fulcrum at the lumbosacral area is the fulcrum that classical dressage has always been most concerned with. It is not the fulcrum the German Military riding establishment was most concerned with.

When Steinbrecht says:

> The borderline between military riding and upper level dressage, if such a line is to be drawn, is defined in that military riding has as its goal to put the horse in balance, that is, to put equal weight on the forehand and the hindquarters, while in upper level dressage the movements become the more perfect the more the center of gravity of the weight can be moved backwards without interfering with the freedom of the hindquarters.
>
> (125)

This observation is not only coming from an expert in upper-level dressage, it is coming from a contemporary of the riders teaching this other, new system. I think it is pretty clear that Steinbrecht saw these two kinds of riding as so different that the one could never be a threat to the tenets of classical dressage.

There was always lurking under the surface a darker side of long and low: that the technique might have more to do with psychological submission from the horse rather than some progressive gymnastic exercise. Biel's observation in the 1800s about German riders riding horses in contorted neck positions and Seeger's warnings to the German riders about Baucher's neck manipulations and harsh techniques were the red flags of the time. I'm not sure anyone knew then how classical dressage was being challenged.

In the 1600s, the Duke of Newcastle William Cavendish said the whole object in schooling the horse is to get the horse upon the haunches. When the hind end comes down and under and the topline is firmly connected, the rider will feel the lift in the back of the horse. It is as if the iliopsoas and psoas system positions the hips and hind end, and then the muscles of the neck, back, and hind legs pull the weight of the front end up over the fulcrum of our seesaw. Just as the seesaw must be rigid, there is a general contraction of the topline of the horse that stiffens to pull against the abdominals and planted hind legs. When the whole ligament system, including the great nuchal ligament, is engaged, it supports the whole topline and gives the muscles room to work. The neck is now firmly woven into the back in a strong ascending arch like a stallion approaching a mare. The contracting abdominal muscles work like a bow string supporting the upward curve of the topline and the horse becomes one piece, firm.

Many riders who have not felt or have been instructed on the feel of true collection like a levade, think this contraction or firmness is wrong or unsettling, so they immediately drop the horse's head and neck down to stretch or relieve the tension. If you constantly loosen the strings of your guitar, you will have no notes to make music. The constant refrain of instructors in this school is to "put your hands down, lower the horse's neck and ride it over the top." The hind end immediately switches from the possibility of lift and carrying power, to thrust and push, overburdening the front end. This kind of riding inadvertently trains the horse to look and move more like a draft horse pulling a log than a dressage horse preparing to levade.

When the rider/trainer can make transitions without the horse dropping its head and neck down or using the reins for support and balance, but instead connects with the hind end, moving the balance back, then eventually in this system the horse will be able to carry the rider in increasing collection that could ultimately result in correct piaffe or even airs. This process is elegantly illustrated in Hans Handler's book, *The Spanish Riding School: Four Centuries of Classical Horsemanship* (Figure 2.6). His simple set of drawings depicts the training of a dressage horse for the last 500 years. We can see the classical road map through the phrases of collecting the horse.

What we have been talking about up until now is the model that has been a consistent template for classical dressage since the 1500s. It culminated from the philosophical requirements of how to ethically ride a horse and through a desire to create a partnership. It is very difficult to overstate how unique this partnership is. In the history of this earth, there has never been a partnership between two different species that have formed a relationship beyond work to create art.

> Once there was a time, in a city, in the kingdom of Naples, known as Sybaris, where not only men but also horses, learned to dance to the sound of the symphony.
>
> (Grisone 65)

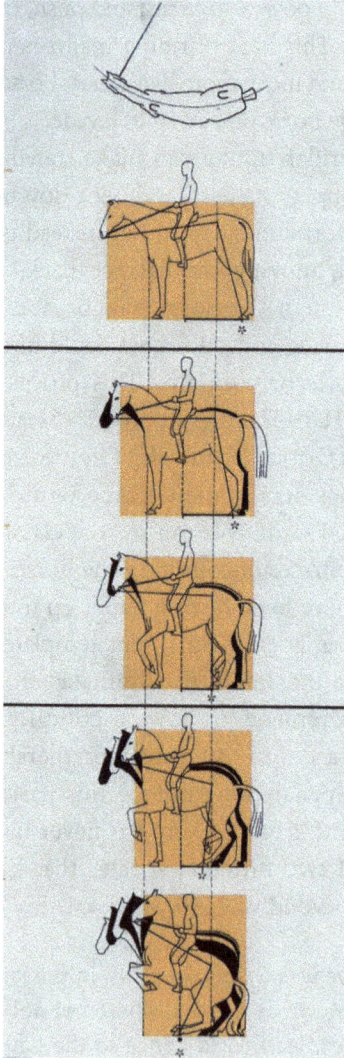

Figure 2.6 The dressage training process is elegantly illustrated in Hans Handler's book, *The Spanish Riding School: Four Centuries of Classical Horsemanship*. Used with permission.

The Second Fulcrum

In the second illustration, we see a different fulcrum. Here the fulcrum is at or near the withers of the horse. In its simplest explanation, when the long neck of the horse reaches down to a place below the withers, the back behind the withers gets lifted up. In our first illustration, riders knew if the horse's head was up and the neck was in a crested position like a stallion approaching a more, the topline could be connected. The neck would be woven into the back and with this firmness they could access the fulcrum near the sacrum. This cresting gesture is a complicated but very symphonic action of principally the splenius muscle which is mainly responsible for the beautiful arch of the horse's neck. It connects with the nuchal ligament, back, and brachiocephalicus which lifts the front leg forward. Their symphonic contractions brace the back against the support of the abdominal muscles. We saw how this contraction is at the heart of the mechanics to lighten the forehand by shifting the horse more upon its haunches.

Now, when the horse is encouraged to lower the neck, in a sense all this supportive bracing must relax, but it is not as if the whole neck just rests. Another set of muscles on the underside of the neck connect from the ribs and sternum to the bottom of the jaw. They are the muscles that pull the horse's head down and nose toward the chest. (If the horse's head and neck is pulled in severely, the horse cannot swallow.) When the head and neck lowers, and the nuchal ligament system gets stretched, it draws on the long processes of the spine near the withers. All of these have a seesaw effect where the rider feels lift in the back (Figure 2.7).

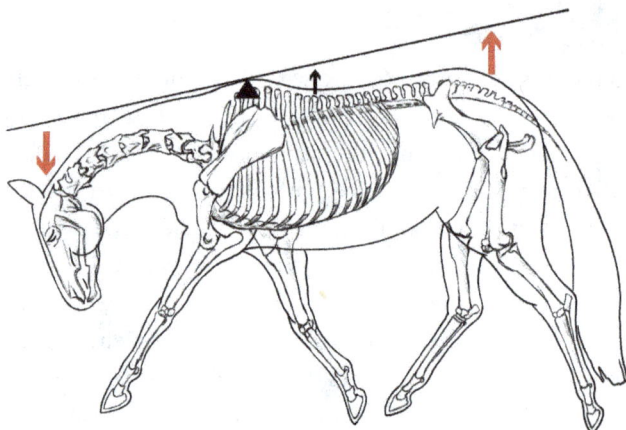

Figure 2.7 In this illustration, the fulcrum is near the withers. As the horse extends its neck out and down like a seesaw, the back goes up. Herein lies the seduction: in long and low, the back will come up under the rider and the rider may think 'how can this be wrong?'. However, the unintended consequence is that the hind end is also lightened and is not developing any musculature for carrying weight and power for collection.

Herein lies the seduction of long and low. It can feel, to an uneducated rider, that there is this positive stretching. The connection and lift the rider feels is from the ligamentous cable system that stretch the entire length of the horse. In long and low, the topline is unsupported by braced muscles of the back and abdominals. The job of holding the topline together is left mainly to the nuchal ligament which draws against the long spinal processes at the withers. The major consequence for dressage training is that this kind of movement lightens the hind end of the horse *in the reverse* of how collection works. Instead of the iliopsoas group tipping the haunches under and then the neck contracting to connect the topline for lift of the front end, the muscles on the underside pull the head and neck down, stretching the topline ligaments to now lighten the hind end, rocking the weight forward. The gesture becomes the antithesis of classic collection. Its use was based on a balancing theory that pretty much exclusively concentrated on the horse's neck, whether it was low "searching for truffles," whether it was then raised up to be in balance, or whether it was flexed and manipulated to weaken it. In this nouvelle dressage, this military dressage, the natural imbalance of weight on the forehand is never really altered, in fact it is made much worse. The unintended consequence of even mild long and low riding would not truly manifest itself for a long time.

3

POSTWAR LONG AND LOW

ON THE FRONT LINES

In the early part of the 1900s, horses were being bred for utilitarian purposes of agriculture and to supply the world's cavalries. The East Prussia horses had to pass simple work-related performance tests. Celia Clarke and Debbie Wallin, in their book *The International Warmblood Horse*, quote Fritz Schilke from his book, *Trakhener Horses, Then and Now*, as he explained the test East Prussian horses were required to perform:

> A. Performance test with a plough, team of two, single share plough, time require 4 hours, pulling

DOI: 10.1201/9781003627579-4

resistance per horse 120 kg, width of furrow 30.35 cm. Minimum area to be ploughed 2.2 morgen (app. 1.4 acres).

B. Performance test in heavy harness: pulling a load, team of two, on hard road for 25 km with a load of 2.5 t including wagon, excluding driver. The first 21 km walk and trot as desired with a maximum time of 6 minutes per km. The last 4 km walk, maximum time 11 minutes per km.

C. Final test under a rider, free walk, free trot without minimum performance. Gallop 2 km in a total maximum time of 5 minutes, 30 seconds.

(Clarke and Wallin, 21–22)

These objectives quickly changed after the world wars when the industrialization of the world made these uses of the horses obsolete. It was then the objectives changed to producing a leisure horse. The German breeding program took off; by the late 1950s and early 1960s, German horses were appearing in Argentina, Venezuela, Mexico, Canada, England, Italy, Sweden, and the United States. Warmbloods in general were appearing everywhere, following the custom of naming the particular subset of warmblood by the location they were from, be it a province or country. Germany was still producing more warmbloods than anywhere else. There was a lot of cross-pollination. Three of the top five German stallions mid-century were French warmbloods.

In the case of the new sport of dressage (competition dressage), the movement of horses began to be a premier factor in judging. A growing fascination with hyper-mobility, especially of the front end, became a craze, including the addition of harness horses into breeding programs to accentuate knee action. With the increasing flamboyance, fundamentals were overlooked and faults in disposition were forgiven. Movement became of utmost importance. Movement was more important than rider position, more important than proper execution of required exercises. It was the number one factor in success in the competition ring aside from politics.

The Changing Horses

I once had a brief conversation when I was a young trainer, a little frustrated at the difficulty in handling some of the newer warmblood horses being bred, with the great German horseman Manfred Lopp who was once the president of the German Hanoverian Stallion Licensing Commission. When I asked him what did he feel was going on, in a very diplomatic way he looked at me and said, "breeders need to ride their horses." I got the message. Breeders needed to prioritize trainability and not sensationalism. He had some the best riders in the world, and even some of them were having trouble managing these horses.

Yet, the objectives of breeding are almost universal in the breed descriptions of any kind of warmblood horse, and they sound something like this:

> The conformation should meet with the demands made of a riding horse. His character and spirit makes him eager to work, co-operative, balanced,

happy, friendly, intelligent and game. The emphasis is on performance, particularly dressage and
jumping, but without forsaking the fundamental
riding horse qualities of conformation, paces and
character.

(Clarke and Wallin, 152)

Who was going to argue with this? Everyone wanted
to buy that kind of horse. However, the breed reality
was drifting from these breed ideals. Even in the 1980s,
when Reiner Klimke was becoming world renowned
for his successes in dressage, he talked about riding his
horses in the beginning of their rides in the long and
low frame because they needed a stretch from being
confined in a stall 24/7 as was the custom in most of
Central Europe. Within a couple of decades of breeding, international dressage competition riders talked
about having horses kept outside 24/7 because they
were too hot to be confined and then safely ridden.
Interviews with successful international trainers talked
about how they sought hot horses for upper-level competition. Combine these leading examples with the
advances in frozen semen and artificial insemination
available worldwide, and anyone who had the money
could breed to the dream stallion they saw advertised
on the internet. The result was the warmbloods at the
end of the 1900s were not the warmbloods at the beginning of the 1900s. This temperament change was going
to play heavily into the rationalization of long and low
that was coming in the 1990s.

On the Front Lines: Teaching Long and Low, but Something Is Wrong with the Syllabus

With the continuing expansion of dressage after the wars and more civilians participating, some began receiving instruction and coaching from trainers of a classical background. People like George Wahl, Franz Rockowansky, and Ernst Bachinger (all from the Spanish Riding School) trained successful competitors. In France, Michel Henriquet and his wife Catherine Durand brought some of the first Iberian horses to world-class competitions. It seemed the military system of long and low was calming down. The German classical teacher, trainer, and icon Egon von Neindorff still used long and low in warming up and cooling down, as did European and Olympic Champion Reiner Klimke, but both always cautioned to use it with a soft rein.

Nevertheless, long and low's association with competitive dressage began to be institutionally embedded in the education of judges. It would take time, but a study would show that judges consistently scored horses higher who were presented behind the vertical (Kienapfel et al., 2024).

Long and low would be incorporated into dressage riding organizations and their educational programs.

There were trainers, including myself, who had experience in different schools and training systems, were still impressed by the classical masters' work, yet thought we saw some possibilities in long and low that could be combined with the classical systems. We tried incorporating it into our training and teaching.

For those of us who were teaching a lot of lessons, even in different countries, problems with long and low quickly began to become more obvious. Teaching more inexperienced riders to put their horse's heads and necks down, especially at the beginning of lessons or workouts, invited disaster. There were often big, strong horses, and the riders were almost begging their horses to act out by putting their horses in the exact posture to buck or play, which they often did. Some trainers thought the answer was to have the student lunge the horse before riding. However, the horse was still in the same posture with long and low side reins, or even no side reins at all, with the intention of getting the bucks out of the horse before they rode. This is an equally disastrous choice. Again, it almost encourages the horse to misbehave and the training equipment should never be associated with play or bad behavior. It is doubly poor judgment because you are usually bringing the horse out from the stable or somewhere where the movement was pretty sedate or have a young horse that is not very strong yet and in a matter of minutes you allow explosive behavior on the end of the lunge line on a relatively small circle. The first strategy put the rider at risk of injury, the second strategy put the horse at risk of injury. When I would go to a clinic and see and talk to new riders, I couldn't believe the list of serious rider injuries that seemed to be acceptable. The long and low strategy was also putting even more weight on the horses already overburdened front end, and this was supposed to be introductory dressage.

Another problem that consistently occurred was that if the riders were taught to get the head and neck of the horses lower, they couldn't get the horses' heads back up. Inexperienced riders didn't have the core strength to block the downward plunge of the horse and at the same time they were not allowed to pull on the reins. They had no tools to try to solve this problem. To add insult to injury, they were often given the same advice as did the German Cavalry Manual H. Dv. 12: "if the horse lowers its head slightly too much, meaning it leans on the bit . . . the rider must energetically ride the horse forward," which only eliminated any chance of developing any carrying power to rebalance the horse (42). In fact, this increased thrust would drive more weight onto the forehand. This made no sense to new riders who had to live with this inner frustration. A standard side effect of this kind of training was the purchase of all kinds of saddles, medical remedies, and finally another horse. For me personally, the most disheartening thing was to watch the love these riders had for riding and their horse dimming. Unless these riders lived in a very horsey area, more often than not they had no choice for different instruction. There were no options, no second opinions to the parochial German competition system, which came from the head-quarters of dressage organizations.

If the riders and horses were a bit more advanced or if the rider were lucky enough to have a horse that had good natural balance, they were experimenting with collection. They might be able to get their horses' heads and necks up, but the horse had been getting no preparatory exercises to teach it how to use its hips to come under the mass, to balance off the forehand. The psoas system of these horses was untrained. When they halted, the horses' hind legs were almost always camped out behind. Their backs were even stiffer than normal because they had been doing a healthy dose of long and low riding, which was tensing the whole topline's ligamentous system, stretching it right into weakening the muscles of the topline. The seductive lift riders were feeling during long and low was often caused by the stretched ligaments, the swing riders were feeling was a result of the muscles of the topline letting go, de-contracting, conceding to the muscles of the underside of the neck that were contracting to pull the neck down and head low. The longer the one stayed in these exercises, the stiffer and weaker the backs and topline actually became. It was obvious from an analysis of many photographs of proponents of the system that horses in upper-level movements were often hollow and not sitting, specifically when compared to riders of the classical school.

The front end of the horse can rise in two ways. One is to increase the ground reaction force: for a force that impacts the ground (the hoof), there will be an equal force back up. If a rider had the horse already on the forehand by riding the neck long and low, and they increased the power by driving the horse more forward, they could get higher action in the front end. The more the force the rider can put on the forelegs, the more they could bounce the forehand up. This lift, of course, has nothing to do with collection. The other way the horse can raise the forehand is to pull the forehand up with a coordinated effort of the muscles of the back and hind end, leveraging the weight back over the hind legs, lightening the forehand.

When these riders who had the horse on the forehand tried to engage their horses with more seat and leg, their driving aids usually only pushed the horse into hovering in the trot, a kind of false passage that was relying more on the ricochet effect of ground reaction force on the legs (rather than more lift from the elastic flexion and extension of powerfully trained muscles). When they used the driving aids of the seat and back and the horse's back and core was weak, the rider pushed the back down and pressurized the stifle and hocks. The bow string of the abdominals gave way and the whole topline sagged. No matter how active the hindquarters were, any lift of the front end would be hollow.

If the riders and horses were working toward the highest levels, Grand Prix movements and/or High School, the shortfalls of the long and low system really showed up. In all fairness to the initial inventors of the long and low system, they might argue that it was never meant to produce or train upper-level dressage horses, that those problems are not their fault. Except that in all physical activities, good fundamentals don't change. If the fundamentals you are teaching put a ceiling on the student's (horse or rider) progress, then their value has to be suspect. Of course, the fundamentals of long and low were pretty much the opposite of the fundamentals of classical dressage, which were a mastery of collection that could lead at its highest level to the Grand Prix movements and airs above the ground.

In the case of the piaffe, for example, often the rider would have a really game horse who would try very hard for the rider, but it had not been strengthened enough behind to help lighten the forehand. In most cases, the horse was accustomed to loading the front legs. The horse would try by pushing the front end up with its front legs, each front leg would step out to the side. As the horse pushed off the right leg, for example, it would push the front end up but also push it hard to the left. The left front leg would then sort of catch the energy and push it back up but hard to the right, starting this tennis match of energy in the front limbs. In other cases, again with insufficient core strength and inability to come under with the hind end to lift and carry, the horse in piaffe would start reaching the front legs back with subsequent steps. The front legs creep back to help lift or at least carry the mass by moving backward closer to the center of mass. The old trainers would call this triangulation, "the elephant balancing on a stool." In many of these attempts at piaffe, the hind legs would actually move closer together, instead of widening their stance to carry more weight. Looking at the horse in profile, it might look like it was engaging the hind legs very well, but from the back one could see the horse was carrying no weight on hind legs, the supposed engagement was a mask with all the carrying being done with the front end again. Trainers would press the horses harder but without the proper fundamental body movements, they would only make it worse.

During passage, the backs would often drop and the hocks would be out behind an imaginary line drawn from the point of the buttocks to the ground. Other times, if the horses didn't have sufficient core strength, they would develop an axial roll in the suspension phase, twisting left and right with each step. They were not strong enough to hold their bodies still, legs landing under the center of mass on a single track.

Canter pirouettes would spin, again as the shoulders tried to do the work of the hind end. There was no end to the complexity of the technical problems that kept occurring. The thing was that these creative evasions (or solutions) the horses were coming up with were all centered in a certain problem. The horse wasn't strong enough to sit and change the weight distribution with correct form. This inability would be tipped off by unusual leg placement, often only slight in the beginning but if the trainer wasn't aware of it and didn't try to correct it, the deviations would become more and more difficult to correct. It seemed to be a vicious cycle of more and more exotic theories of correction. Yet no matter how complex the evasion became, and how difficult and complex the correction could be, the cause was often just the opposite.

Suppose I go to the gym and work on simple curls with dumbbells. My trainer notices I'm getting tired and I'm not using my arms correctly. I'm arching my back, bending backward, jerking the weight up. He stops me and we try to fix the form, or maybe rest or simply have to wait until the next day to recover. Too many dressage training stables were looking like a bad gym, full of all kinds of contortion with coaches not correcting poor technique and improper fundamentals.

It was very simple, you couldn't wait until the upper levels to teach collection. Young horses had to learn the posture early. The more you kept a horse on the forehand

and off the haunches, the more difficult collection would become. In terms of rider education and participation, there was an obvious bottleneck few riders could get past at a very elementary level. The ceiling was not their effort as much as it was the system of teaching.

What made the job more difficult for many of us trainers was that horses trained this way were being rewarded and were winning in the 1970s and 1980s. Horses carried their heads too low or were behind the bit and placed higher than horses who were on the bit or even slightly above. Collection was less important than rhythm or movement. Horses trained in the military long and low system were winning, but it wasn't because of superior performances or technique. It was because there was shift in the paradigm of what upper-level dressage should now look like.

Much of the struggle to embody this new dressage was hidden from the public eye. You had to somehow be on the inside to see what was actually being done, what the training techniques were, and by whom. Draw reins and bitting rigs became common equipment even in supposedly the most prestigious stables. Pully systems were attached to legs. You had to be careful of criticism. There were powerful people in the burgeoning dressage industry. The rewarding of this continuous obsession with horse's necks and submission was encouraging the acceptance of more and more forms of aggressive aversion training. Classical dressage was not a compliment. It was bound to get worse if something did not change. It was hard to see where the change might come from.

In this new sport, there was sophisticated mutually beneficial partnership between judges, breeders, coaches, agents of sales, competitors, and officers of organizations. These jobs were often interchangeable. A judge for example could also be a competitor, coach of a competitor, breeder, seller, or vice versa, a situation ripe for conflict of interest. In this mutually incentivizing system, the paradigm of military riding came embedded.

The irony is that horse breeders should know the dangers of inbreeding. For many of us, we seemed to be living in a real-life enactment of Juan Manel's and Hans Christian Anderson's fable of the emperor's new (no) clothes. A group of people with power convince the king that the garment they have made for him can only be seen by men of legitimate birth, or in Anderson's story, by clever, competent, special people. A relatively small but powerful clique had manufactured a reality, but their collective denial of its effects could only last so long.

The interesting thing about the sport of dressage is that in the initial set of rules, there was really nothing that was not classical. Sport dressage didn't need to change dressage, but it did.

Somehow the riders of long and low had convinced themselves this method was superior to classical dressage. It was almost like some article of faith that couldn't be altered by any amount of evidence. The leaders of this church were often charismatic and later they had the powerful rewards of competitions to reinforce their teaching: rewards of team selections, of notoriety, of prizes. Their teachings could not explain away the problems we were seeing on the ground. This situation was putting teachers and trainers to a test. Which system were they going to promote?

4

1980–1990

THE GLOVES COME OFF

Rollkur

There were many rumors swirling. The current star of the German competitive dressage scene was a young woman named Nicole Uphoff, and her horse was named Rembrandt. She had been training with Uwe Schulten-Baumer since 1986; a year later she won her first Grand Prix at Lausanne. Four months before the Seoul Olympics, after some personal issues she switched trainers to Harry

DOI: 10.1201/9781003627579-5

Boldt, a trainer and coach, and solid member of the German establishment. The judges were loving this new phenom. However, people were seeing a disturbing style of riding when she was practicing and warming up. She would pull the head and neck of her horse down and bend it from one side to another, with the horse's nose close to its chest. It was severe hand riding. When it was noticed, she denied it as a routine correction. She even told people that the horse liked a good stretch. The press would not have it. St. Georg magazine, an institution in equestrian journalism, published a photo of her riding Rembrandt, schooling. It was as severe as the rumors. The public started to become incensed. In February 1994, her coach Harry Boldt was asked about it for the German magazine Reiter Revue. "It was an extreme exception, Rembrandt shied for fences erected at the end of the arena. The fact that Nicole curled him down so extraordinarily deep was only for correctional purposes." It was the same denial, "it was only a momentary correction."

Around the same time, an American videographer named Beth Baumert had been given permission to film the warm-ups at the Aachen dressage show, a famous German venue for horse shows. She filmed many riders and included interviews in a series of videos she made. She has said she had no intention of making anyone look bad or preparing some exposé. As a dressage rider and teacher herself, she said she "just wanted to see what the Europeans were doing" (Baumert, 2023). However, in one of the films, there is clear sustained footage of Uphoff riding Rembrandt in the extreme long and low flexions for long periods of time. It was clearly not a momentary correction, it was a well-practiced continuous technique. To add to the evidence, there was more video of her riding another horse, Grand Gilbert. Uphoff seems to be getting exasperated, and riding him with the same technique. The horse's neck and head are pulled in and held in for long periods of time, then switching bend.

There was now irrefutable proof. It wasn't possible for Uphoff to deny it any longer. The public became increasingly incensed. St. Georg magazine named the new version of long and low, "rollkur," a derogatory expression from the German "Rollkur" which was a stomach remedy that people would take and then roll on the floor for it to coat the stomach. If the public's wrath was unrelenting, the judges' interpretation only seemed to be more resolved (Figure 4.1).

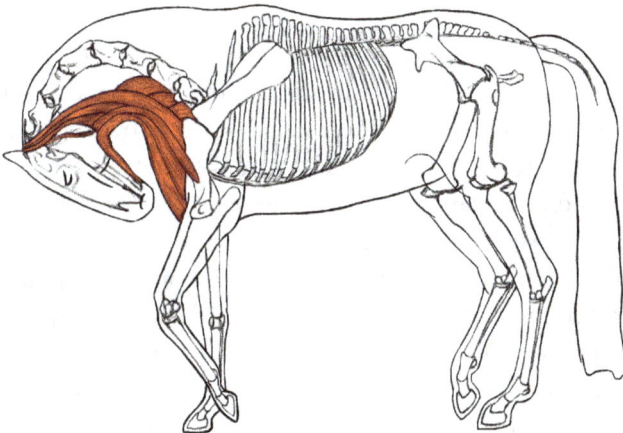

Figure 4.1 Hyperflexion: "The neck muscles are being trained in an undesirable way," said McGreavy, "you're strengthening the underline, not the topline."

Wolfgang Niggli was an engineer by trade. Earlier in his career, he was an officer in the Swiss Cavalry. As a rider, he rode mainly jumpers and competed in steeplechases. He was not noted to be a dressage rider. Niggli was the chairman of the FEI Dressage Committee during 1981–1993. As the head FEI judge, he said Nicole Uphoff and Rembrandt's performances "moved him to tears" (St. Georg). FEI Judge Nick Williams said, Rembrandt "epitomized impulsion and lightness" (Horse and Hound, 2001). While the upper echelons of sport dressage were in collective denial of what was going on in the sport, the public was not. However, to blame the abusive techniques on one 20-year-old woman was not only unfair, it was untrue: Uphoff had said more than once that Uwe Schulten-Baumer gave her the key as to how to ride Rembrandt.

Uwe Schulten-Baumer

Uwe Schulten-Baumer was the son of a farmer. He had a humble beginning, where he groomed in trade for his first riding lessons. While he was in the military, he rode his Commandant's horse on weekends. He rode jumpers and showed at Aachen. In the world of jumpers, it was common to use creative and severe bits, draw reins, and controlling devices. The jumper riders wanted quick responses and at the higher levels, complete obedience. The good ones trained their horses, the ambitious ones sought quick fixes and didn't rule out brutality.

It is interesting that if you read the current edition of *The German Riding and Driving System Book 2*, The Official Instruction Handbook of the German National Equestrian

Federation: Advanced Techniques of Riding," which is a modern equivalent of the H. Dv. 12 Army Regulation 12 German Cavalry Manual of Training of Horse and Rider, you see very classical language. They claim in Section 2, The Training of the Horse and Rider in Advanced Dressage, that the German system is based mainly on the teachings of Guérinière, which is exactly what Podhajsky had to say about the teaching and training at the Spanish Riding School.

In fact, much of the language that follows sounds like Steinbrecht. There is no mention of long and low that was the center of training in much of Germany and the standard in sport dressage. All references to dropping the head down seem to be carefully scrubbed. The only place losgellaissenhiet is mentioned is in the section on training the jumper:

> But in contrast to the dressage horse, the show jumpers work in a deeper outline. He does not have to have the same degree of collection, therefore he cannot—or does not need to—have the same elevated outline. . . . Only horses with extremely difficult conformation . . . may justify temporary use of running reins.
>
> (99–100)

The reality was different.

Schulten-Baumer came to dressage not really as a rider of dressage. He worked as a manager in the steel and cement industry. He had no time to compete personally, but gained his reputation from coaching dressage riders, including his own children. Schulten-Baumer's son and daughter competed at the highest levels of dressage. While his son was studying for his medical career, he kept his horse at Warendorff where he was said to have daily lessons with Willie Schulthies, the student of Otto Lörke, Gustav Rau's early architect of the School of Hanover, and the military-influenced competition system.

It seems only logical that Schulten-Baumer would bring his background in jumping to sport dressage. Schulten-Baumer's methods as a coach came to the international forefront when he began coaching Nicole Uphoff and Rembrandt in 1986. Rollkur, the severe hand riding, was a technique which had no secondary or tertiary value. It was a tool for submission, period. When the horse would totally yield to the rider's hands, when the rider could dominate the neck and put it anywhere the rider wanted, then the rider could dominate the resigned horse and make it obey.

The claims that this technique was horse-specific to Rembrandt due to his personality lost all credibility when the same year in 1986 Schulten-Baumer began training Isabell Werth on one of his own horses, Gigolo. Again, there were videos showing almost the exact neck position as with Nicole Uphoff and Rembrandt. Werth is bending Gigolo's head down, nose close to the chest, the neck bent strongly to the left and right, the hind legs are camped out and unengaged to his croup-high bouncing. The hand riding is severe and relentless until the horse yields.

When the horse's head is pulled in severely, the topline muscles must let go as the muscles on the underside of the neck, the sternomandibillares, the sternohyoidaris, lugus collis, and the tongue contract to the point where the horse is unable to even swallow. The process works the way "submission holds" in jujitsu do: if applied correctly, the opponent gives up.

One of the great risks of this type of training is it can easily devolve into a learned helplessness, which is when an individual continuously faces a negative, uncontrollable situation and stops trying to change their circumstances even when they have the ability to do so. In 1965, Martin Seligman was studying classical conditioning. The experimenters rang a bell and then administered a mild electric shock to a dog. After only a few tries, the bell would ring and the dog would respond as if it had been shocked. There was a second part of the experiment. They built a box divided in half by a small fence. One side had an electrical floor and the other side was normal. If they shocked a normal dog, it would jump the fence to avoid the shock. If they used one of the dogs that been conditioned in the previous experiment, the dog wouldn't jump away. It would simply lie down in passive resignation.

At the time of rollkur, there were no significant scientific studies to prove the effects of this kind of severe negative reinforcement or aversion training in riding. It seemed simply to be a study in submission which would equal compliance, the compliance would complement the desired performance. The obvious abuse was repeatedly overlooked, continuing a cycle of more abuse. Nicole Uphoff and Rembrandt continued winning. The judges who were rewarding her performances and others using similarly abusive training systems always had the same rationalizations. They only had authority to make judgments on what happened in the show ring in front of them. If the riders brought the horses' heads up for the test, that was all they could address. They couldn't address the brutality of warm-ups, or the training systems. That might have worked, except that the way the horses were being trained did show up in the performances. Horses trained with all this emphasis on the forehand often showed strong faults in collection, which is where judges could have helped with a correction, of course. There was an opportunity for education. In Rembrandt's passage, the horse marks time but was consistently out behind in the haunches with a hollow back. His piaffe never sat. All the areas of collection were faulty, yet they received very high marks.

The same faults were obvious in the performances of Gigolo and others. The unintended consequences, by now almost 100 years, of long and low training was that horses trained that way were losing the core strength and ability to collect. Judges were unable to see it or unwilling to recognize its integral importance in evaluating correct dressage training.

The trend toward rewarding hypermobility of the front end and high-energy performances kept growing. Younger judges were indoctrinated in this new view, imitating the scoring of senior judges and being trained on a model of these newer (and always warmblood) performances. Not all competitors were following this trend, there were examples of horses and riders who were showing more classical or traditional training, but they were not winning. Two factions in dressage riders were galvanizing. Classicists saw the brutality as both a philosophical descension of dressage back to the days before Pluvinel and physically a decline in the quality of performances. The revolutionaries promoted this new style and defended it as a way to govern the horse while trying to elicit maximum effort in their performances.

If the public had bought into the long and low military system, they were reaching a limit. For the FEI, it was a different story. By the time Nicole Uphoff retired Rembrandt, the FEI judges had rewarded her with at least 25 international medals, 24 of them gold. By the time Isabell Werth retired Gigolo, FEI judges had rewarded her with at least 22 international medals (usually gold, one silver). The competition between Nicole Uphoff and Isabell Werth was soon eclipsed by Isabell Werth's rivalry with another young woman, Dutch rider Anky van Grunsven on a horse named Bonfire. It seems as if the Dutch had been watching the success of the Germans with long and low riding, and were going to make their own mark.

5

1900S–2000S

A GIRAFFE IS A GIRAFFE, NO MATTER WHAT YOU WANT TO CALL IT

It seemed like deja vu. Another young, attractive woman riding a powerful warmblood horse in almost manic forward-charging performances, the same curled-in neck practices, the same rewards from the judges, the same fawning comments from video announcers. It was in the 2000s that the public was being captivated by social media platforms and the power of its voice. The outcry about the brutality of this new version of long and low continued to grow in strength. The FEI was not budging on addressing

DOI: 10.1201/9781003627579-6

the brutality from more and more aggressive hand riding and head and neck manipulations of the horse.

If the Germans were mostly denying the practice or were embarrassed by it, the Dutch were going to take a new tack, that the best defense was a good offense. Anky van Grunsven partnered with a man named Sjef Janssen who became her coach and later her husband, and subsequently coach of the Dutch dressage team. Sjef Janssen did not start riding until he was 28 years old. He, like so many of the promoters of long and low, did not have an education in classical dressage. He had limited success as a rider, but primarily came into his own as coach of his future wife and then the Dutch dressage team. Anky Van Grunsven rode from childhood. Neither of the two were shrinking violets. When the public and press criticized their deep riding practices, they pushed back hard with threats and lawsuits while they rebranded rollkur. They claimed they were not practicing rollkur, but were practicing long, deep, and round (LDR.) Photos and films showed virtually no difference in the postures of their horses and previous photos and films of horses in rollkur. It was the same curled up neck, the severe bending of the neck from one side and the other. Their claim was not that they were not doing that—they freely admitted it and went on to champion what they were doing as a new system to train dressage. Sjef Janssen said:

> It is a totally different approach because I don't come out of the horse world. I come from a background where sport is very important. . . . [T]he

> system is based on taking away the blocks in the horse. . . . Rule number one is that the horse always thinks forward. . . . So, if a horse is thinking a little backwards, we solved the problem by getting them extremely forward.

When asked more about neck position, he said "We can also ride very short, but we can also ride them immediately extremely long. They always follow the hand wherever we want them" (Hector, 2014).

If that seemed to be an echo of Francois Baucher or the Hanover School, that's because it was.

The goal of Baucher's method was the total submission of the will of the horse to the will of the horseman at any moment.

In *Francois Baucher, The Man and His Method*, scholar and translator Hilda Nelson writes:

> When the horse determines his own strength, Baucher calls this strength *instinctive force*. If the horseman determines the horse's strength, he calls this strength *transmitted force*. In the first case, man is dominated by the horse; he becomes the victim of the horse's will and caprices. In the second instance, the horseman makes a docile instrument out of the horse and submits him to all the impulsions of his will. Once a horse is mounted, he must only function in accordance with transmitted strength. The constant application of this principle determines the talent and expertise of the horseman. To succeed in totally dominating the horse, the horseman must dominate the horse's strength.

(At first, Baucher said that the horseman must destroy this strength. After 1842, détruites became réduites, that is reduced.) It is evident that the basis of Boucher's new method is the elimination, destruction or reduction of the *instinctive* strength of the horse and its replacement by the *transmitted* strength that emanates from the horseman.

(22–23)

In terms of the German Hanover School, Schwabl von Gordon writes emphatically: "Following any activity in a working frame or in a dressage frame, the horse should willingly and immediately reply with the request: 'From the outside regulating rein, have the horse stretch into the long and low position!'" (18).

Whatever you think about van Grunsven and Janssen's arguments and practices, somehow they were persuasive enough to convince the FEI that their system was exempt. In 2010, facing mounting pressure to do something, the FEI held a roundtable conference at the IOC headquarters at Lausanne to address the crisis. The FEI President HRH Haya accepted a petition from Dr. Gerd Heuschmann of 40,000 signatures against rollkur. At this same conference, Frank Kemperman from Holland was the Dressage Committee Chair and Sjef Janssen was a dressage representative. The FEI said they had resolved the controversy, and summarized their meeting with this statement:

The consensus of the group is that any head and neck position of the horse achieved through aggressive force is not acceptable. The group redefined hyperflexion/Rollkur as flexion of the horse's

> neck achieved through aggressive force, which is
> therefore unacceptable. The technique known as
> Long, Deep and Round (LDR) which achieves flex-
> ion without undue force is acceptable.
>
> (FEI.org, 2010)

Apparently having the horses in extreme positions of
flexion was ok with the FEI. Their objection seemed to
be with how aggressive the force was that it took to get
them there.

Additionally, it was decided:

> The FEI will establish a working group, headed by
> Dressage Committee Chair Frank Kemperman, to
> expand the current guidelines for stewards to facili-
> tate the implementation of this policy. The group
> agreed that no changes are required to the current
> FEI Rules.
>
> (FEI.org, 2010)

By then the head and neck riding had spread everywhere. In
2009, the British Horse Society demanded an inquiry into
the Swedish rider Patrick Kittel for warming up his horse
Watermill Scandic in sustained positions of hyperflexion. "During the first part of the training session . . . [t]he
tongue was clearly blue, and flopped limply from the horse's
mouth" (Dressage-News.com, 2009). The FEI legal depart-
ment issued no formal claim against Kittel: "In a statement,
the International Equestrian Federation says studies of video
evidence and witness statements concluded that there is 'no
reliable evidence that the warm-up techniques used by Mr
Kittel were excessive'" (Horse and Hound.com, 2010).

Once again, the press and public were outraged. St *Georg* magazine published an article "Dressage Perverse" at the time. In 2015, the Danish Equestrian Federation found rider Andreas Helgstrand guilty of "improper use of the bit and bridle," after a Danish TV station released pictures of his horse with a blue tongue caused by lack of blood supply (Heath, 2015).

In 2017, seven years after the FEI's roundtable meeting in Lausanne where the FEI decided to establish a working group headed by Dressage Committee Chair Frank Kemperman, the FEI stated:

> There are no known clinical side effects specifically arising from the use of hyperflexion, however there are serious concerns for a horse's well-being if the technique is not practiced correctly. The FEI considers hyperflexion in any equestrian sport as an example of mental abuse. The FEI states it does not support the practice.
>
> (Jurga, 2017)

There was no way to read the above paragraph other than the FEI did not have serious concerns if the technique was "practiced correctly." Someone had persuaded the FEI that the mental abuse of hyperflexion could be practiced correctly.

By now some serious new research had begun to come out, looking past the disguises of these pseudo-stretching regimens as training systems for the horse's body to whether they were in fact more simple aversion training systems based on operant conditioning principles using strong negative reinforcement. One of the most relevant studies was of different training methods of dogs, conducted during 2016–2019 and published in the National Institute of Health's National Library of Medicine.

> To our knowledge this is the first comprehensive and systematic study to evaluate and report the effects of dog training methods on companion dog welfare. Critically, our study points to the fact that the welfare of companion dogs trained with aversive-based methods is at risk, especially if these are used in high proportions.
>
> (Castro et al., 2020)

The study pointed to higher cortisol loads after these training sessions and increased nervous behaviors, and so on.

In horses, Uta Koenig von Borstel, PhD, BSc, a professor at Germany's University of Gottingen, carried out a review of existing scientific literature. "The team looked at 55 scientific articles dealing with the effects of head and neck position (HNP) on a various types of horses' welfare and/or performance" (Beckstett, 2015). Coauthor Paul McGreevy, BVSc, PhD, MRCVS, MACVS (Animal Welfare) Cert CABC, an animal behavior and welfare science professor at the University of Sydney, presented the results on von Borstel's behalf at the International Society of Equitation Science Conference.

> Of those studies, 88% indicated that a hyper flexed HNP negatively impacts welfare via airway obstruction, pathological changes in the neck structure, impaired forward vision, stress and pain. . . . Based

on this review, wrote Koenig von Borstel, "the pre-
sumed gymnastic benefits of training horses in a
hyper flexed head and neck position are by far out-
weighed by both undesired gymnastic effects and
reduced equine welfare."

(Beckstett, 2015)

Additionally,

23% of studies suggested [hyperflexion] has negative influences, such as . . . a reduced oxygen supply
and increased activation of the lower neck muscles.
"The neck muscles are being trained in an undesirable way," said McGreevy. You're strengthening the
underline, not the topline.

(Beckstett, 2015)

This lead us back to the very beginning, the difference
between classical and nouvelle dressage. Classicists felt it
was only ethical to prepare the horse to carry weight by
developing the hindquarters and topline and rebalancing
the horse more toward the rear, whereas nouvelle dressage riders tip the horse on the forehand by lowering the
head and neck.

Dr. Hilary Clayton, BVMS, DACVSMR, PhD, professor emerita at Michigan State University, consultant on
horsesport science, renowned biomechanics expert, and
Grand Prix dressage rider/trainer herself, said: "Pre-
exercise muscle stretches are not recommended for
horses who must produce high levels of muscle strength
or power . . . holding an unnatural position for a pro-
longed period of time is not good for the horse" (Pascoe,

2023). She advocates that trainers work more on core strength instead.

There continues to be more and more evidence on detrimental effects of extreme long and low regimens. In spite of all this evidence, the FEI's responses continue to be mysterious. From a period of time from about 2014 to 2023, there had been a steady scrubbing of the dressage rules. Whereas before there were clear definitions and requirements for horses in competitions, that is, being on the bit with the poll at the highest point, with the face at or in front of vertical in all movements, they are gone. It seems inconceivable that judges who are supposed to rule on law will have no laws to guide them. Perhaps that was the point, to eliminate the classical definitions in order to make way for nouvelle dressage definitions.

It is very difficult to understand the actions of the FEI. On one hand, you have the mission statement of the FEI where they use language like, "horse first," "athlete welfare," "sustainability," "happy horses," "fair and equal," "education," and at the same time in the period of time roughly from the 1990s into the 2000s, three riders dominated the world's attention on rollkur. In spite of the public outcry and the considerable evidence, Nicole Uphoff riding for Germany, Isabell Werth riding for Germany, and Anky von Grunsven riding for Holland all were awarded well over 100 international medals by FEI judges at different Olympics, European Championships, and World Championships, most of which were gold and silver. The actions of the FEI speak more loudly than their words. How would aspiring young dressage enthusiasts and their coaches look at this legacy of the paragons of dressage sport and not copy them? Once again, long and low proved its inextricable connection to winning at dressage sport competitions.

6

"A PARALLEL UNIVERSE"

Before the beginning of long and low, before the beginning of sport dressage, there was another kind of dressage being practiced throughout Europe. Perhaps nowhere in the world could you see a better real-life demonstration of this kind of dressage than at a performance of the Spanish Riding School in Vienna, Austria. A north star for the Spanish Riding School was Francois de la Guérinière who was inspired by a predecessor, Antoine de Pluvinel. Their philosophy and technical skill were in turn inspired

DOI: 10.1201/9781003627579-7

by Xenophon's famous axiom that "Anything forced and misunderstood can never be beautiful" (62).

In her translation of Pluvinel's *Le Maneige Royal*, Hilda Nelson explains: "It is Xenophon's book, *On Equitation*, that came into the hands of the Neapolitan nobleman Frederico Grisone, known to his contemporaries as the father of the art of equitation and director of the famous riding school of Naples." In *Gli Ordini di Cavalcare*, published in 1550, Grisone followed the training techniques for horse and rider as they had been set down by Xenophon. However, he ignored, for the most part, Xenophon's ideas on the use of "gentling" techniques during the training of the young horse. Xenophon's notions of good horsemanship (as expressed not only in his *The Art of Horsemanship*, but also in his other works) were primarily based on treating the horse with kindness—on "gentling" rather than "breaking the horse." "Anything forced and misunderstood," says Xenophon, "can never be beautiful." This is the fundamental notion of classical beauty: beauty based on harmony, order, moderation, and the following of nature; the very notions that the men of the 17th century—and with them Pluvinel—stressed (de Pluvinel, viii).

The Spanish Riding School's reliance on Guérinière's work is described nicely here:

> [Francois Robichon, Seiur del la Guérinière] was regarded by his contemporaries as the first authority on horseback riding. In 1716, he opened a riding academy in Versailles which soon became famous throughout Europe. From 1730 until his death on

July 2, 1751, he was Master of the Horses at the Court of Louis the XV of France. Following in Pluvinel's footsteps, he emphatically rejected all brutal dressage methods and introduced quite a few improvements in the High School of Horsemanship, including the oblique "shoulder in" movement. His *Ecole de Cavalerie, contenant la connaissance, l'instruction et la conservation du cheval* (School of Horsemanship, Including Information on Horses, Their Training and Their Care) put the entire art of riding on a new scientific basis which has proved of great use to many generations. The system laid down by La Guérineire in this "Ecole" was a determining factor also for the "Spanische Reitschule Wien." This school, which is the last stronghold of classical horsemanship, will continue, at least where its basic principles are concerned, to utilize these fundamentals.

(The Spanish Riding School, 9)

Inside the jackets of the riders of the SRS were sewn sugar pockets. Their humble uniforms in earthy brown showed no marking of rank, their whips were modest tree branches. They were not supposed to be the center of attention. It was all carefully designed to accent humility not ego, to let the training speak for itself, to showcase the horses and a training system that was passed down pretty much orally for hundreds of years. There was no more complete explanation of the definitions and standards of classical dressage than one of the their famous performances, a combination of live theater, ballet, opera, documentary, lesson, lecture. The choreography of the production, like the acts of a great play, was masterful in itself. The contents of each act were staggering.

The Performance

In the first act, a group of 5-year-old stallions, still dappled or dark gray like immature swans, come into the exquisite Spanish Riding School. The hall is an architectural treasure in Vienna, with its marbled tiers and vaulted ceiling that seems like it is a mile above the floor. The young stallions walk, trot, and canter; there is no fixed routine. The riders give each other room, and the horses are ridden forward in an uphill frame. They have been started in a lunging program, and are now under saddle beginning the phase of strengthening and learning control. They change rein to check symmetry, they make simple transitions to teach obedience. As an audience you are shown the very beginning, you get to pick your favorites, you wonder how will each one turn out. It almost feels like you are a part of the performance.

In the next act, you experience nothing short of time travel. The next group of horses that enter are pure white, mature swans. These horses and riders will exhibit the highest levels of work on the ground. They show lengthening, collection, piaffe, passage, canter pirouettes, flying changes, three tempis, two tempis, and finally straight changes every single stride. The riders sit still in a posture they have been practicing since they were very young. You hardly know which horse to watch, it is easy to be overwhelmed. On to the third act where two horses and riders are carefully matched to perform a Pas De Deux, a mirror performance at times in perfect symmetry, understruck by Mozart's Symphony No. 40 in G Minor. The movements are so similar they seem to echo in the great hall like two great singers' harmonic notes in perfect pitch.

In the fourth act we get to see behind the scenes, the magicians show you how the illusions are accomplished. Here the transparency of this training system is completely revealed, there are no tricks. The riders work the horses in hand, unmounted. There is just a single line to the bridle in one hand, and in the other a whip. They guide the horse, teaching it to increase and control its emotions. They deepen the set of the haunches, they perfect the piaffe, controlling it to make a reliable platform for the explosive caprioles, the supernatural jumps of the courbette, the subtle and most exquisite control of power in the levade, where the horse is balanced motionless on two hind legs for seconds that seem like hours as a horse and rider turn into an ethereal piece of sculpture before your eyes. There has to be so much trust and skill to practice such movements that are filled with risk.

You see how much practice goes into this ignition of forces. A touch of fire to light a firework that reveals itself in the air in a different form like a blooming flower or a finished painting, drifting in the sky only long enough for your eye to recognize it. The horse softly lands, the rider holding the line in one hand is already reaching behind his back for a piece of sugar before the horse has even stopped moving. If there are mistakes, there is no aversion training. They will adjust and try again. If things don't go well, there is not more punishment, there is more practice. No one is going to pull the horse's head down until its will is broken or it rebels.

You begin to wonder—what are they really showing you here? Are these tricks like at a circus or a metaphor of something much more philosophical, like how to practice? Riders wear the same clothes to practice as they do in performance. They bring the same seriousness and dignity. How they get the result is almost more important than the result. Deliberately controlling the highest emotions, calmly standing next to a mature stallion— a 1,000-pound animal who is as fit as any elite athlete about to attempt something at the limits of its body. It is a meditation as serious as the ancient samurai. The Zen master Shunryu Suzuki told his students, "It is easy to have calmness in inactivity, it is hard to have calmness in activity, but calmness in activity is true calmness."

The program continues with the fifth act. Almost as if to give the audience relief from the adrenaline of the last act, the music changes and marches by Walch, CM Ziekner, J. Fucik, set the scene for a single horse being guided by one rider walking behind it holding two long reins. The horse performs half passes, piaffes, passages, even canter and canter pirouettes, flying changes. You wonder how it is possible for the horse to change gaits and do all these movements when all the while the rider never leaves his steady walk.

The sixth act is the ultimate in high-school dressage training. All of the airs we saw being trained unmounted will now actually be ridden without stirrups. There seems to be even more tension as the riders glance at each other, to see if a runway is clear. Like some busy airport, they will need room to prepare for the flights and may have to wait or circle. You watch the engravings of Pluvinel and Guérinière come to life, and need to realize the rarity of what you are seeing. Even among the Spanish Riding School stallions, only a select few come up every few years who are capable of this pinnacle of mastery.

The program ends with the famous quadrille, a ballet with as many as 12 horses set to the music of Chopin, Bizet, and others. They perform in precision, so close together they are almost on top of one another. Sliding together and apart like a desk of cards in a magician's hand, they fan out and reconvene. They perform multiple pirouettes simultaneously and from above it looks like petals of a flower unfolding and folding back again. It is a grand finish, yet right before it there will be a solo performance, the penultimate act in this revelatory opera.

In 1994, when rollkur was the rage and the public was in outcry from violent hand riding, Klaus Krzisch rode Siglavy Mantua. It was a gala performance in honor of the restoration of the SRS after a catastrophic fire. The school chandeliers seemed even brighter in the freshly refinished school. Krzisch rode in as he had many times before. Slowly, he took off his hat to salute the large portrait of Emperor Charles VI. He put his hat back on, settled it, then raised the point of his whip up into the air with his right hand. He would not need it anymore. He took up the curb rein alone in his left hand, and placed the hand over the mane of the stallion. It would not move through the entire performance. In perfect position with legs quiet, they began. As if they were reading each other's minds, they performed everything. Trot half passes, counter changes of hand with no kicking legs, no apparent steering. Trot to piaffe, piaffe back to trot, turning corners, perfectly threading the narrow pillars, piaffe pirouettes, passage. They were so linked, it was a real demonstration of the mythical classical seat that riders study and study for years to learn. Minutes kept going by, and the difficulty only increased, steadily accompanied by the waltzes of Vienna.

On to the canter. Krzisch rode three tempis, two tempis, one tempis down the centerline, perfectly straight through the narrow pillars. The pair collected seamlessly into a canter pirouette, not one revolution but three. Then it is back onto the straight centerline, immediately into more one tempis, then another collection to another precise pirouette. They seemed tireless. They finished on a long piece of passage in a perfect, infectious rhythm. The audience was dying to clap along but they contained themselves, they had too much respect. The pair humbly finished. There was no waving to the audience. It was an exquisite, almost exhausting display of the highest level of dressage that has been evolving for almost 2,000 years. With his hat at his side, he disappeared through the back entrance.

There is an expression they used in the training of the SRS riders. It was to learn to become "the thinking rider." When the competition dressage judges argue that they can only judge what is in front of them, that is nothing but an excuse. When younger trainers and riders copy long and low systems to make money or try to gain fame, there is no excuse. For a thinking rider, or a thinking judge, or a thinking student of dressage, or even a thinking audience, there is always an accessible alternative to inhumanity and an assault on nature, to the uncontrolled desires of the ego. A horseman cannot say in good conscience they didn't know what was going on. Xenophon has been gone for 2,000 years, Guérinière has been gone for hundreds of years. Even if the Spanish Riding School ceases to exist, its influence remains. It is easily available. You cannot lead an authentic life and say I can only judge what is in front of me, casting a blind eye to your peripheral vision. Today more than ever, we need new Xenophons, Pluvinels, Guérinières. Not more riders chasing prizes and money (Figure 6.1).

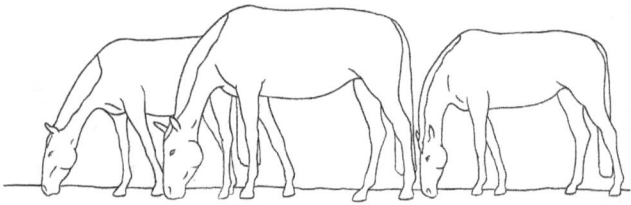

Figure 6.1 All by themselves, horses are masters of long and low.

CONCLUSION

Within days after a foal is born, even though it needs its mother's milk to survive, it will begin "searching for truffles," nibbling bits of hay or grass. By the time the horse is 3 years old and is ready to begin training, it has spent half its life in a long and low position grazing, drinking, inspecting. It can twist its head in the lowest, contorted positions to get a choice piece of grass under a fence. Why would a training system think the young horse needs more of this? It is by now a self-made master

DOI: 10.1201/9781003627579-8

of long and low. Long and low as a stretching regimen has always been suspect. From a training point of view, what made Baron Biel nervous, what made Podhajsky nervous, what made the classical horsemen nervous was not the stretching, it was the shadow side of long and low. It was the School of Hanover's insistence on immediate lowering of the horse's head with the action of the outside rein. It was Baucher's neck and jaw manipulations to rob the horse of all its power. It was the systematic evolution of increasingly more and more aggressive hand riding.

It was a consequential drift away from sugar pockets toward the strictest aversion training. The exhibitions in sport dressage became more demonstrations of control of the horse rather than demonstrations of cooperation with the horse. Biel, Podhajsky, Seeger, and many others were standing on the edge of a seismic paradigm shift that was every bit as frightening as standing on the edge of an active earthquake fault. The paradigm shift was a new definition of dressage which would move away from the classical dressage methods practiced at the Spanish Riding School in Austria, at Samur in France, and in Spain and Portugal which were based on the mastery of longitudinal and lateral balance, and a love and admiration for the horse, to a nouvelle dressage of utility practiced in the militaries which created and governed the new sport of dressage. For the military minds, there didn't seem to be any reason why this art form couldn't be calibrated like jumping and military (now ironically the original name for three-day eventing) to select a winner. As a rule, for education in the arts societies, don't look to the military where following orders and strict, complete obedience are the heart and soul of the mindset. But they did here.

There were probably two significant factors which fueled the long and low movement. One was the formation of the rules for dressage competitions, where airs above the ground were not allowed or were eliminated. Testing would stop at piaffe, passage, pirouettes, and so on. It was a little like stopping an education before university. The unintended consequence was that it would open the door to the shift of fulcrums in the sport of dressage. You would not need the mastery of collection, even if it were just a philosophical homage to the airs. You would be able to get away with piaffes that didn't sit, that were not platforms for the airs. The engagement of the haunches, the holy grail of the Duke of Newcastle, Guérienière, Pluvinel, et al. lost out to the balance more on the forehand of Baucher and the German military. The second force in effect was the burgeoning sport horse industry. If Gustav Rau had a dream, it would have been to put everyone on a riding horse and that riding horse had better be a German warmblood. Remember, Rau tried to get Hanoverians to replace Lipizanners in the Spanish Riding School. Fritz Schilke, the father of the Trakhener breed, warned whoever would listen that no breed ever died from lack of type, but from lack of marketability. From the beginning, Rau was out front in the promotion of the German sport horse and breeding. The Germans were brilliant in the promotion of their horses. In terms of dressage, it was important that their horses win at the new competitions. It would be very good for business. There was no reason why other breeds could not participate in sport dressage. There would have to be consolidated effort to associate the German warmblood with winning, and there was.

From the beginning, the sport of dressage was consumed with politics and intrigue, to the point where after the Stockholm Olympics in 1952 the IOC was going to eliminate dressage from the Olympics. Remember the German judges deliberately adjusting Podhajsky's score down in order ensure the German win in 1936? That kind of politics just continued. After a promise of reform from Prince Bernhard of the Netherlands, the IOC relented.

However, the burgeoning sport horse breeding industry was going to ensure that the next generations of judging candidates were indoctrinated. They would learn to judge dressage by watching primarily warmblood horses and memorizing the German military training scale. These programs were so successful that a group of scientists in peer-reviewed study finally proved in the 2000s that an implicit bias was so strong it ensured a significant advantage to the warmbloods. The paradigm shift was complete. Gustav Rau would have been very proud.

Yet in a strange way, once a certain level of homogeneity had been achieved in dressage shows, breeders had bred themselves into a corner. If 20 riders rode into a class and they all were on warmbloods, how would the winner be determined? Of course, political biases were embedded in the DNA of sport dressage, but there was another element of selecting a winner and it was movement which became decisive. In a race to breed a more and more forward horse with more and more flamboy-ant (hyper) movement, sometimes hyper-dispositions and hyper-medical problems came along with it. Excuses for the "necessary" hyper-aversion training of hyperflex-ions followed. It was all unsustainable. But the business was booming, with record prices for stallions and sales of top prospects. No one seemed to know how to turn it around. Along with the pressures to produce winning horses, scandals continued and only grew. It seemed it would only be a matter of time before the IOC would have to step in again. The sport of dressage could not control itself. The question for the next generation of riders was going to be, "what kind of dressage do you want to practice?" It seemed one generation of riders had already been lost to the worst of long and low. But, there always had been a choice.

My old friend Sheila McLevedge who was Henri van Schaik's assistant for many years, reminded me of van Schaik's story of a man who wanted to take Egon von Neindorff to Warendorf to see the best of the upcom-ing young German riders. Von Neindorff went. He sat through three riders, then he stood up and left. His host was startled. He followed von Neindorff out. "What's the matter?" he asked. Von Neindorff looked at him with a sad expression and said, "It's a shame they don't like horses."

BIBLIOGRAPHY

Appels, Astrid. "Anky Van Grunsven Sues Eurodressage." *Eurodressage.* www.eurodressage.com/2010/08/25/anky-van-grunsven-sues-eurodressage. August 25, 2010.

Baumert, Beth. Personal interview. November 20, 2023.

Beckstett, Alexandra. "Hyperflexion in Review." *The Horse. com.* https://thehorse.com/113453/hyperflexion-in-review. November 5, 2015.

Belasik, Paul. *Dressage for the 21st Century.* Trafalgar Square Publishing, 2001.

Belasik, Paul. *Dressage for No Country.* Trafalgar Square Publishing, 2019.

Belasik, Paul. *Exploring Dressage Technique: Journeys into the Art of Classical Riding.* 1994. J. A. Allen and Co. Ltd., 1998.

Clarke, Celia, and Wallin, DebbieDebbie. *The International Warmblood Horse.* Kenilworth Press Ltd., 1991.

Clayton, Hilary. *The Dynamic Horse*. Sport Horse Publications, 2004.

De Pluvinel, Antoine. *The Maneige Royal*. 1626. *Translated by Hilda Nelson*. J. A. Allen and Co. Ltd., 1977.

Diggle, Martin. *The Illustrated Encyclopedia of Dressage*. Trafalgar Square Publishing, 2005.

"Dressage Star Rembrandt Dies." *Horse and Hound*. www.horseandhound.co.uk/dressage/dressage-star-rembrandt-dies-38243. November 6, 2001.

"FEI Launches Investigation of Sweden's Patrick Kittel Warming up Horse at Odense World Cup." *Dressage-News.com*. www.dressage-news.com/2009/10/27/feu-launches-investigation-of-swedens-patrik-kittel-warming-up-horse-at-odense-world-cup. October 27, 2009.

"FEI Round Table Conference Resolves Rollkur Controversy." *FEI.org*. https://inside.fei.org/media-updates/fei-round-table-conference-resolves-rollkür-controversy. February 8, 2010.

German National Equestrian Federation. *The German Riding and Driving System, Book 2, Advanced Techniques of Riding*. 2nd ed. *Translated by Gisela Holstein*. Kenilworth Press Ltd., 1996.

Grisone, Federico. *The Rules of Riding*. 1550. Translated by Elizabeth M. Tobey and Federica Deigan. Arizona Center for Medieval and Renaissance Studies, 2014.

Handler, Hans. *The Spanish Riding School: Four Centuries of Classic Horsemanship*. McGraw-Hill Book Co., 1972.

H. Dv. 12: Army Riding Regulation 12: German Cavalry Manual on the Training of the Horse and Rider. 1937. *Translated by Stefanie Reinhold*. Xenophon Press, 2014.

Heath, Sophia. "Leading Dressage Rider Guilty of 'Improper Use of Bit and Bridle'." *Horse and Hound.com*. www.

horseandhound.co.uk/news/andreas-helgstrand-guilty-improper-use-bit-bridle-blue-tongue-473300. February 4, 2015.

Hector, Chris. "Sjef Janssen Explains His Training System." *The Horse Magazine.* www.horsemagazine.com/thm/2014/12/sjef-janssen-explains-his-training-system. December 3, 2014.

Heuschman, Dr. Gerd. *Tug of War: Classical Versus 'Modern' Dressage.* 2006. Translated by Reina Abelhauser. Trafalgar Square, 2007.

Jurga, Fran. "Rollkur Revolt: FEI Makes Official Statement Discouraging Overflexion in Dressage Training." *Equus Magazine.* www.equusmagazine.com/blog-equus/rollkur-revolt-fei-makes-official-statement-discouraging-overflexion-in-dressage-training. March 10, 2017.

Kienapfel, K., Piccolo, L., Cockburn, M., Gmel, A., Rueb, D., and Bachmann, I. "Comparison of Head-Neck Positions and Conflict Behavior in Ridden Elite Dressage Horses between Warm-Up and Competition." *Applied Animal Behavior Science,* 272, 2024, 106202. www.sciencedirect.com/sciencearticle/pii/S0168159124000509.

Letts, Elizabeth. *The Perfect Horse.* Ballantine Books, 2016.

Nelson, Hilda. *Francois Baucher: The Man and His Method.* J. A. Allen and Co., 1992.

Pascoe, Elaine. "Hyperflexion: Going to Extremes." *Practical Horseman.* www.practicalhorsemanmag.com/health/hyper-flexion-in-horses. April 9, 2023.

"Patrick Kittel Escapes Disciplinary Action by FEI over Blue Tongue." *Horse and Hound.com.* www.horseandhound.co.uk/news/patrick-kittel-escapes-disciplinary-action-by-fei-over-blue-tongue-294320. January 27, 2010.

Pickerel, Tamsin. *The Horse: 30,000 Years of the Horse in Art*. Merrill, 2006.

Podhajsky, Alois. *The Art of Dressage: Basic Principles of Riding and Judging*. 1974. *Translated by Eva Podhajsky*. Doubleday and Company, Inc., 1976.

Podhajsky, Alois. *My Dancing White Horses*. 1960. Translated by Frances Hogarth-Gaute. Holt, Rinehart and Winston, 1965.

Schilke, Dr. Fritz. *Trakehner Horses, Then and Now*. 1974. *Translated by Helen K. Gibble*. American Trakehner Association, Inc., 1977.

Schwabl von Gordon, Gert. *Classical Dressage Training in Practice, According to the H. Dv. 12*. Xenophon Verlag, 2022.

Seeger, Louis. *Monsieur Baucher and His Art: A Serious Word with Germany's Riders*. 1852. *Translated by Cynthia Hodges*. Auriga Books, 2010.

Seunig, Waldemar. *Horsemanship: A Comprehensive Book and Training the Horse and Its Rider*. Translated by Leonard Mins. *Doubleday and Co*. Inc., 1965.

The Spanish Riding School. Forum Verlag, 1985.

Spanish Riding School of Vienna. Anglo-Austrian Society, Freedman Bros. Ltd., 1993.

Steinbrecht, Gustav. *The Gymnasium of the Horse*. German 10th ed. *1978*. Translated by Helen K. Gibble. Xenophon Press, 1995.

USDF Lungeing Manual. United States Dressage Federation, rev. 2008.

Vieira de Castro, Ana Catarina, Fuchs, Danielle, Morello, Gabriela Munhoz, Pastur, Stefania, de Sousa, Liliana, and Olsson, Anna S. "Does Training Method Matter? Evidence for the Negative Impact of Aversive-Based Methods on Companion Dog Welfare." *National Library of Medicine*, 2020. www.ncbi.nlm.nih.gov/pmc/articles/PMC7743949.

Werth, Isabell, and Simeoni, EliEli. *Four Legs Move My Soul: The Authorized Biography of Dressage Olympian Isabell Werth*. 2018. Translated by Lena Reinderman. Trafalgar Square, 2019.

Wyche, Sara. *The Horse's Muscles in Motion*. The Crowood Press Ltd., 2002.

Wyche, Sara. *Understanding the Horse's Back*. The Crowood Press Ltd., 1998.

Xenophon. *The Art of Horsemanship*. J. A. Allen, 1979.

For Product Safety Concerns and Information please contact our EU
representative GPSR@taylorandfrancis.com
Taylor & Francis Verlag GmbH, Kaufingerstraße 24, 80331 München, Germany

www.ingramcontent.com/pod-product-compliance
Lightning Source LLC
Chambersburg PA
CBHW050540270326
41926CB00015B/3311